Lucas Vinícius Reis
André Luiz Emmel Silva
Jorge André Ribas Moraes

Proposal to implement P+L in a metalworking company

Lucas Vinícius Reis
André Luiz Emmel Silva
Jorge André Ribas Moraes

Proposal to implement P+L in a metalworking company

Ways to integrate productivity and sustainability

ScienciaScripts

Imprint

Cover image: www.ingimage.com

This book is a translation from the original published under ISBN 978-3-330-76644-0.

Publisher:
Sciencia Scripts
is a trademark of
Dodo Books Indian Ocean Ltd. and OmniScriptum S.R.L publishing group

120 High Road, East Finchley, London, N2 9ED, United Kingdom
Str. Armeneasca 28/1, office 1, Chisinau MD-2012, Republic of Moldova, Europe
Managing Directors: Ieva Konstantinova, Victoria Ursu
info@omniscriptum.com

Printed at: see last page
ISBN: 978-620-8-54031-9

SUMMARY

The growing demand for sustainable products and the increased rigour of environmental legislation in most countries has forced companies to adopt less polluting means of production, introducing this variable into their strategic planning. The aim of this study is to develop a Cleaner Production implementation proposal for a small company in the metal-mechanics sector, following the implementation methodology suggested by the National Centre for Clean Technologies - CNTL. It was classified as exploratory research, using a quantitative approach and the case study method as the collection procedure. Thus, the barriers to implementation were identified, the production process flow chart was developed with its inputs and outputs quantified and the causes of waste generation were established. Using this data, a proposal for environmental performance indicators was developed to monitor the results of implementing this methodology and, finally, P+L opportunities were proposed according to the priority level established by the CNTL, totalling six level 1 actions and one level 3 action. At the end of the study it was possible to conclude that P+L is a viable option for the industrial sector studied, offering opportunities to eliminate waste at the source, as well as economic advantages and improving the company's image.

Key words: Cleaner production, Metal-mechanics industry, Environmental indicators, Industrial waste.

SUMMARY

SUBJECT AREA AND LIMITATIONS

This work was developed in the area of Sustainability Engineering, more specifically on the subject of Cleaner Production (P+L). It is limited to a proposal for the application of P+L in the production process of the Metal Cruz company, exclusively in the product line supplied to customer Xalingo, from June 2014 to June 2015.

CHAPTER 1

INTRODUCTION

In this chapter, the research will be presented through the history of the company, the justification that led to its realisation and, finally, the general and specific objectives established for the study.

1.1 Company history

Lair da Cruz was founded in 2010 and is located at 210 Capão da Cruz Road in the Progresso neighbourhood of Santa Cruz do Sul. It operates in the metal-mechanical sector, where it is better known by its trade name Metal Cruz, and provides turning services to a variety of companies in a wide range of industrial sectors. The owner and founder, Lair da Cruz, started the company's activities in the current $108m^2$ building (figure 1), where he installed a lathe and two bench drills and began producing small spare parts for the Xalingo company, as well as carrying out small services for mechanical workshops, clothing manufacturers and the general public. In 2011, Metal Cruz acquired welding machines, a transfer press, a hydraulic press and a saw for cutting steel and metal billets, increasing the range of services offered.

Figure 1 - Metal Cruz façade

Source: Company archive, 2015.

Metal Cruz currently has one employee working a single shift, and has Xalingo as its main customer, for whom it supplies a range of products.

12 products, including clamps for rotomoulding and spare parts for these clamps (Figure 2).

Figure 2 - Metal Cruz's main products

Source: Company archive, 2015.

The company also works on adapting machinery to NBR 12 standards, providing industrial maintenance services and making metal parts and structures for companies in a wide range of industries. In order to meet demand, Metal Cruz outsources some processes.

1.2 Justification

Sustainable products and processes, free from environmental liabilities, are objectives pursued by almost all companies. This awareness began to take shape in the last decades of the 20th century, and has been disseminated around the world through studies, reports, treaties and conferences (DIAS, 2008), such as the United Nations Conference on the Human Environment held in Stockholm in 1972, the Rio92 conference in 1992, the Kyoto Protocol in 1997, the Copenhagen conference in 2008 and countless other events that are recognised worldwide .

These events drew the world's attention to the need to adopt policies and methodologies aimed at environmental preservation and prompted a series of scientific studies, which according to Pereira, Silva and Carbonari (2012) show that the balance of ecosystems is at risk in various parts of the planet, which could lead to a shortage of the natural resources needed to sustain human life on Earth, in line with the economic model that has prevailed since the industrial revolution in the 18th century.

> Numerous organisations have developed and implemented various environmental management philosophies, practices and techniques to meet market demands. These include: waste management, natural and energy waste management, environmental management systems, labelling, environmental labels, certification, eco-efficiency systems, cleaner production, eco-design, life cycle analysis, reverse logistics systems, green supply chain management, among others (AMATO NETO, 2011).

Among these methodologies aimed at facilitating the process of implementing sustainable economic models by the industrial and service sectors, Cleaner Production (P+L) has found satisfactory results in various sectors of the economy in Brazil and abroad.

P+L is a tool developed by UNEP (United Nations Environment Programme) and applied as an alternative to replace traditional control models, known as "end of *pipe*", which disregard the causes of waste generation and act only at the final stage of the process (DIAS, 2011; LIMA et al., 2014b).

This methodology has proven to be effective in reducing environmental impacts in various sectors of industry, but it is rarely applied in Brazil, mainly due to a lack of knowledge and studies about it. It could be of great value in the metal-mechanics sector, as this is a sector that generates a lot of contaminated waste and has a high environmental impact.

> The metalworking industry processes different types of materials such as steel, iron, copper, plastic, wood, paints and solvents. It also has activities and processes with a high potential for environmental contamination, such as electroplating, machining, heat treatment and surface treatment (DIAS, 2008).

By analysing the process of implementing the P+L methodology in an industrial sector that is a major player in the country's economy and which also has a high risk of generating environmental liabilities, this study can make a significant contribution to the development of a theoretical framework for this area of research. Although this is a case study whose application is focussed on the production process of the company studied, the description of the stages of application of P+L and its results can serve as a basis for other companies in the sector.

Quantifying the inputs and outputs of the production process and identifying opportunities to reduce waste generation provides environmental and financial gains that can be monitored using appropriate indicators, making it clear to the company the importance of environmental awareness and driving future P+L actions. This quantification also serves to confirm the advantages of applying the methodology to small companies.

Another important factor is the possibility of discovering solutions and good production practices that can be disseminated and adapted to other companies, promoting the evolution of production methods in the metal-mechanics sector. The company under study will benefit from the lessons learnt about the P+L methodology and the need for continuous improvement of processes and workflows, a proposal to achieve a production process with less waste generation and indicators to monitor its progress and, above all, the owner's ability to evaluate the production process using performance indicators and develop improvement actions.

1.2.1 P+L in the scientific world

P+L is a subject that is increasingly debated in scientific circles and accepted by companies (CETESB, 2008; OLIVEIRA et al., 2013). In recent years, this methodology has become the subject of publications at the country's main production engineering events. Through searches using the terms "Cleaner Production", "P+L" and "PmaisL" in the two most recognised national events in the field of Production Engineering, Table 1 was drawn up, showing the number of scientific articles related to the P+L methodology published over the last 10 years.

Table 1 - Publications on the Cleaner Production methodology

SIMPEP		ENEGEP	
Year of publication	Number of publications	Year of publication	Number of publications
2014	5	2014	8
2013	2	2013	5
2012	4	2012	11
2011	4	2011	3
2010	5	2010	1
2009	6	2009	7
2008	4	2008	7
2007	3	2007	11
2006	5	2006	1
2005	1	2005	0

The publications shown in table 1 originate mainly from case studies carried out in companies from a wide range of industries, a fact that demonstrates organisations' interest in the methodology and also the contribution that events and scientific articles make to the dissemination of P+L.

1.3 Objectives

1.3.1 General objective

To propose the application of the Cleaner Production methodology in the production process of the Metal Cruz company.

1.3.2 Specific objectives

- Investigate the generation of losses during the production process;
- Describe the steps for applying the P+L methodology;
- Identify P+L opportunities;
- Propose environmental performance indicators capable of monitoring the production process in terms of the actions implemented.

CHAPTER 2

THEORETICAL BACKGROUND

This chapter deals with the scientific basis used to support the study, starting with the concepts and methods applied by experts in the field.

2.1 Cleaner production

The growing pressure of human activities on the environment demands complete and efficient environmental protection systems (ZHANG et al., 2014). Based on this thinking, in 1989 UNEP developed the concepts of P+L - Cleaner Production - and described it as the continuous application of a preventive environmental strategy applied to processes, products and services.

> It is a tool that seeks to minimise losses, waste and emissions by focusing on eliminating the generating factors in the design of products and services, rather than treating them after they have been created. For processes it includes conserving raw materials and energy, eliminating and reducing toxic emissions and waste (UNEP, 1998).

This methodology understands the benefits brought about by the industrial revolution, but seeks a way to minimise the negative impacts on the environment and the lives of current and future generations.

> Throughout the Industrial Revolution, manufactured capital (equipment, machinery, physical infrastructure) was always seen as the main factor in industrial production, while natural capital (natural resources, living systems and ecosystem services: carbon recycling by plants), seen as "marginal inputs", was only taken into account during periods of war, when scarcity became a problem (Hawken et al. 1999).

However, in today's world, the scarcity of natural resources is a reality arising from the industrial revolution itself and this situation demands an attitude from governments, companies and society. It is becoming increasingly clear that thinking about environmental preservation is no longer a differentiator, but a necessity for maintaining life and capitalism itself, as a lack of resources threatens even production processes. The planet's economic model will only be bearable if the natural environment that sustains it is too (BROWN et al., 2000).

Cleaner Production is an increasingly important part of planning, design, operation and management in all industrial sectors (KLEMES, VARBANOV and HUISINGH, 2012). The concept of Cleaner Production, developed over the last two decades, has brought some innovative environmental thinking to the industrial sector (ZHANG et al., 2014), consolidating itself as a company-specific preventive environmental protection initiative aimed at minimising waste and emissions and maximising production (ALI and FRESNER, 2006).

> In this methodology, there is a need for commitment from all hierarchical levels of the company, resulting in constant questioning about the reasons why waste is generated and how to act to eliminate the generating source, thus forming a new mentality within the organisation (SILVA et al., 2014a).

It can be understood as a different way of analysing the company's processes and the waste it generates. The problem to be solved in P+L is not how the waste will be treated or how it will be disposed of, but rather the reasons that generate it and how it will be eliminated.

P+L differs from traditional processes, where control is only carried out at the final stage of the process, also known as the 'end of pipe' *(ennd-ff-pipé)*. Thus, in P+L, control is carried out during the production process and seeks to prevent the generation of contamination at source (BASS, 2007; DIAS, 2011; LIMA et al., 2014b).

Experiences with this methodology have shown positive results, through technological changes in production processes or even in the way these processes are managed. The changes proposed by Cleaner Production start with the design of the product, the choice of materials, equipment and forms of energy to be used and extend throughout the product's life cycle, even considering its useful life and recycling options after disposal. However, it is important to point out that P+L can be introduced gradually within organisations, starting with actions that do not require large financial investments and are, in the majority of cases, focused on production methods.

Hirschhorn (1997) states that P+L practices include changes in management or operating procedures that generate greater efficiency in the use of raw materials, energy and water.

In addition, P+L generates economic advantages for companies by reducing production costs. The advantages are significant for everyone involved, from the individual to society, from the country to the planet (CETESB, 2008). Generating waste means transforming raw materials and inputs into products with little or no added value, or even costing money to treat and dispose of, reducing profits and even making business unviable (COELHO, 2004).

2.1.1 Barriers to implementing Cleaner Production

P+L encounters difficulties in implementation, especially among small and medium-sized companies, due to a series of factors originating in the most varied aspects that influence the organisational environment. Numerous authors have mentioned these barriers in their research (DONG et al, 2012; KITZBERGER, 2012; SILVA et al, 2014b; SHI et al, 2007). Table 1 lists some of the main barriers to implementing P+L, according to their classification.

Table 1 - Classification of P+L implementation barriers

CLASSIFICATION	DESCRIPTION OF THE BARRIERS
Economic	• Lack of financial resources; • Failure to incorporate environmental costs into investment analyses; • Lack of tax incentives;
Systemic	• Lack of employee training; • Lack or failure of environmental documentation; • Inadequate management system;
Organisational	• Lack of employee involvement; • Privileging productivity to the detriment of environmental issues;
Technique	• Technological limitations; • Lack of resources needed for data collection;

Behavioural	• Resistance to change; • **Lack of culture in "best operating practices";**
Government	• **Concentrating efforts on "end-of-pipe" techniques;** • Lack of adequate legislation to promote the minimisation of environmental damage;

Source: Adapted from UNEP (2002).

Kitzberger et al. (2012) consider that the main difficulties in implementing P+L in small companies are mainly related to a lack of awareness of its benefits, little knowledge on the part of managers, failure to recognise environmental damage, lack of financial resources or even human resources with specialised knowledge in this area. However, changes in consumer mentality, associated with changes in environmental legislation regulations and greater access to information on the part of entrepreneurs have eliminated most of these barriers and made the real benefits of P+L clearer (SILVA, 2014b).

2.1.1 Cleaner Production performance indicators

Performance indicators are instruments capable of providing information on a given reality, allowing this information to be analysed and essential aspects to be extracted (MITCHELL, 2004; PEREIRA, SILVA and CARBONARI, 2012). The indicator helps us to understand the organisation's current situation, which path needs to be followed and how far we have to go to reach the target set (CARDOSO, 2004).

The decision-making process is made easier with the use of appropriate indicators (CAMPOS and MELO, 2008; SILVA, 2014a), increasing the assertiveness index and speeding up the management of the company's processes. The P+L methodology has its actions monitored by indicators capable of measuring water and energy consumption, the amount of raw materials consumed and waste generated per unit produced and the organisation's effluent generation, among others that can be established according to the company's type of activity, its size and the characteristics of the production process.

In the context of sustainability, performance indicators must be able to assess, measure and monitor sustainability, taking into account projections of future scenarios and not just observations of the current situation (PEREIRA, SILVA and CARBONARI, 2012).

> Decision-making tools must take into account the associated risks, meeting the economic objectives of the activities, guaranteeing the economic stability of employees, but also promoting actions to protect ecosystems from threats arising from human activities (WU, OLSON and BIRGE, 2013).

This tool can also be used to present stakeholders, including clients and the community, with the results obtained through actions to prevent and control environmental damage.

For Cleaner Production, indicators are fundamental for providing information on managerial and technological aspects, making it possible to measure the economic benefits obtained from environmental improvements (CARDOSO, 2004).

2.1.2 P+L in Brazil

In Brazil, SENAI - the National Industrial Apprenticeship Service - was the body chosen by UNEP to disseminate the concepts of Cleaner Production through the National Centre for Clean Technologies - CNTL. The CNTL is responsible for providing information, offering courses and giving all the necessary support to companies interested in adopting this tool. Other organisations and institutions are also contributing to the dissemination of P+L in Brazil, some of which are listed in Table 2.

Despite the number of organisations geared towards introducing Cleaner Production in the Brazilian industrial sector, the adoption rate is low, mainly due to the lack of awareness of its benefits and the lack of strategies to overcome the barriers to implementing this methodology.

The evolution of environmental issues has led companies to focus on the source of their solid waste, atmospheric emissions and liquid effluents, seeking solutions in their own production processes (CNTL, 2015a). Based on this thinking, the CNTL has developed a three-level structure for Brazilian companies to define P+L, as shown in figure 3.

Chart 2 - Institutions supporting and disseminating the P+L methodology in Brazil

SIGLA	SIGNIFICANCE
ABIPTI	Brazilian Association of Technological Research Institutions
BNDES	National Bank for Economic and Social Development
BNB	Bank of the Northeast
IDB	World Bank
CEBDS	Brazilian Business Council for Sustainable Development
CNI	National Confederation of Industries
CETESB	São Paulo Basic Sanitation Technology Centre
FINEP	Financier of Studies and Projects
MMA	Ministry of the Environment
NPLs	Cleaner Production Centres
SEBRAE	Brazilian Micro and Small Business Support Service
UFBA	Federal University of Bahia

Source: Adapted from Coelho (2004).

This structure is based on the same ideology used in other countries around the world and prioritises level 1 actions, i.e. avoiding the generation of waste and emissions through changes to the process, the product, the raw material or technological changes. For waste that can't be avoided, actions go to level 2, which seeks to reintegrate this waste back into the process. When level 1 and 2 actions are not enough to completely eliminate waste and emissions, external recycling is sought, which is the action corresponding to level 3.

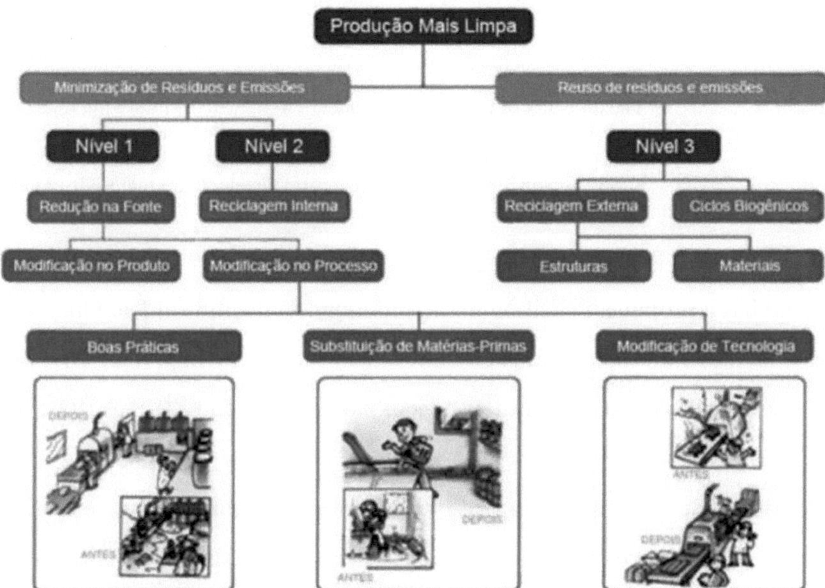

Figure 3 - Flowchart of Cleaner Production levels
Source: Adapted from CNTL (2003).

2.2 Metal-mechanical industries

The metal-mechanics sector has a prominent position in a country's economic development, as it is responsible for developing technologies and acts as a developer of other industrial sectors. It can also be considered a complex and diverse sector, as it ranges from small to large companies and has several different segments. As such, this industrial sector has a large number of companies, with diverse production processes and their own characteristics.

> For Selene et al. (2014), its various segments have a common characteristic: the main component of the goods and services produced is directly related to the production, processing and utilisation of metals, with different levels of technological complexity and a high level of automation.

The development of technologies in this sector boosts other branches of industry, as the majority of companies in the metal-mechanics sector are geared towards the production and commercialisation of machinery and equipment used in production activities in other industrial sectors.

> The diversity of this sector includes activities in the following industrial branches: metallurgical industry, machinery and equipment industry; final goods industry, and other activities such as the production of tools, hardware and other metal artefacts and the electrical materials industry (LIMA et al., 2014a).

In the state of Rio Grande do Sul, the metal-mechanics industry is well developed and has a prominent position in the economy. In the hills of Rio Grande do Sul, there is a metal-mechanics cluster that is considered an important Brazilian industrial centre and is the second largest of its kind in the country (SEVERO et al., 2014).

According to Selene et al. (2014), despite generating technological development, in Brazil this sector operates with outdated machinery and needs a competitive edge and innovation. One way of increasing competitiveness and reducing costs can be achieved through P+L which, according to Severo et al. (2014), is a methodology capable of generating an increase in organisational performance and improving aspects such as flexibility.

These additional advantages to the main focus of P+L can fit perfectly into the metal-mechanics sector, which has a range of high-value raw materials and usually generates a lot of waste during production activities.

2.2.1 Environmental impacts of metal-mechanic production

ISO 14001 (2004) defines environmental impact as any change in the environment, whether adverse or beneficial, that results, in whole or in part, from the organisation's environmental aspects (elements of an organisation's activities, products or services that can interact with the environment).

The metal-mechanics sector uses raw materials and processes that generate a large amount of contaminated waste that is difficult to recycle, especially in machining and milling activities where cutting fluids are the main contaminating agent and therefore generate negative environmental impacts.

> According to Tan et al. (2002), during the machining process, cutting fluid is one of the main causes of environmental pollution, as it causes environmental impacts such as the emission of toxic gases, solid waste and hazardous oily liquid effluents, which can pollute water resources, soil and air.

The high consumption of electricity in this industrial sector is another relevant factor that causes harmful environmental impacts, due to the fact that most companies use automated processes, but equipment that is technologically outdated. The metal-mechanics industry normally uses machines and equipment that, because they don't have a modern electronic system, use an excessive number of motors and usually have an oversized capacity.

> Romm (2004) comments that most of the motors used in industry are inefficient and exceed the required capacity, and that they are usually the most responsible for the company's carbon dioxide emissions. Therefore, modernising processes by implementing control systems and using modern, efficient motors creates opportunities for energy savings.

Companies in the 21st century need to be aware of the most efficient ways of managing their processes in order to reduce the environmental impacts that degrade the environment, adapting to current global environmental norms and standards and, above all, fitting into the production model demanded by today's consumers and the scarcity of resources that has become a reality.

Inter-organisational relations, green marketing, eco-efficiency, among others, are subjects that are being widely addressed, transforming the old business and market vision into a much broader and more complex concept (SILVA et al., 2014b). The adoption of environmental impact management and control tools has been put on the agenda of senior management meetings.

2.3 Waste

In addition to producing products, industrial processes also produce material that is undesirable for the company's purposes and which poses a potential risk to the environment and/or the health of living beings, which is known as waste. When carrying out industrial processes that transform various natural resources into industrialised products, man inevitably generates waste, because according to the law of conservation of mass, matter cannot be totally created or consumed (BRAGA et al., 2002).

2.3.1 Classes of solid waste

Solid waste is divided into two groups, hazardous and non-hazardous, and is classified using the ABNT NBR 10004 standard, which establishes the classification criteria and codes that identify each waste according to its characteristics.

Solid waste is solid and semi-solid waste from industrial, domestic, hospital, commercial, agricultural, service and sweeping activities, including some liquids that cannot be discharged into the sewage system (ABNT - NBR 10004, 2004). There are three classifications for this type of waste: class I waste, class IIA waste and class IIB waste.

Class I corresponds to hazardous waste, which may cause risks to public health or the environment due to its physical, chemical or infectious properties, or may have characteristics of flammability, corrosiveness, reactivity, toxicity or pathogenicity. Class IIA corresponds to the group of non-hazardous, non-inert waste, which may be biodegradable, combustible or soluble in water. Class IIB, on the other hand, represents the group of inert non-hazardous waste, made up of those which, when subjected to static and dynamic contact with distilled or deionised water at room temperature, in accordance with ABNT NBR 10006, do not have any of their constituents solubilised at concentrations above the standards of water potability (ABNT NBR 10004, 2004).

Knowing the classes of solid waste makes it possible for industry to correctly dispose of and treat them, and can also be used to prioritise actions to combat waste generation within the production process.

CHAPTER 3

METHODOLOGY

This chapter covers the classification of the work in terms of its nature, objectives, technical procedures and the way in which the problem was approached. It also describes the scientific methods used and the steps taken to carry out the work.

In terms of objectives, this research can be considered exploratory, as it makes it possible to develop hypotheses, brings the researcher closer to the object of study and allows for the modification of concepts or broadens knowledge of them (MARCONI and LAKATOS, 2008; GIL, 2010), defines the phenomenon studied precisely and promotes knowledge of the relationships between its elements (CERVO, BERVIAN and SILVA 2007). Its approach is quantitative, allowing numerical values to be extracted and analysed (MALHOTRA et al., 2005), and qualitative because the data requires logical treatment by the researcher in order to better apply the methodology (SANTOS, 2000).

As for the collection procedure, the method used is the case study which, according to Yin (2015), allows in-depth knowledge of contemporary events in their real-world context, through direct observation, and may use interviews with people who relate directly to the object of study. As far as sources of information are concerned, this study falls under the heading of field research, since it is an observation of a phenomenon that happens spontaneously, followed by data collection and recording of relevant variables through previously established objectives (MARCONI and LAKATOS, 2008). Figure 4 shows the methodology used in this research in flowchart form, dividing it into stages.

Stage 1: A topic was selected that favours the company's development, benefits society and the environment and brings contributions to the academic community. The objectives to be achieved were established by monitoring the production process, so that an appropriate research structure could be developed based on these.

Stage 2: We searched for theoretical references in books, scientific articles and master's dissertations.

Stage 3: The procedures that guided the work were developed, such as how to identify problems and how to proceed. The data collection techniques were designed to measure all the inputs and outputs of raw materials and energy at each stage of the production process.

Stage 4: The data was collected, analysed and quantified, allowing the causes of waste to be identified and environmental performance indicators to be selected for monitoring the improvements to be implemented and for comparisons to be made with historical data, while also allowing opportunities for improvement to be formulated.

Stage 5: This stage made it possible to assess the effectiveness of the proposal in meeting the objectives and to identify the need to reformulate actions, or even opportunities to carry out new studies.

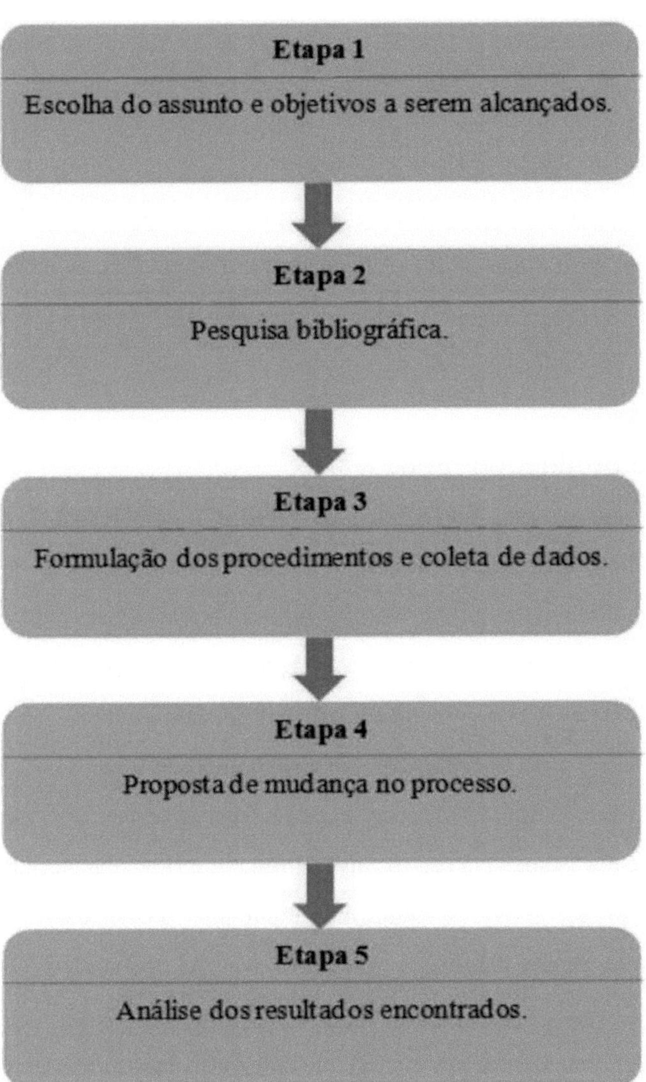

Figure 4 - Flowchart of the research methodology

CHAPTER 4

DEVELOPMENT

This chapter describes the production process under study and also outlines the stages of the P+L implementation process, according to the structure suggested by the CNTL. The aim was to adapt these stages to the characteristics of Metal Cruz in order to achieve the objectives of this case study without changing the scope of the methodology.

4.1 Description of the production process

All of Metal Cruz's raw materials and inputs are stored in a demarcated area inside the company's building, and then processed according to the product to be manufactured, which can go through the cutting, drilling, turning, threading, assembly, welding and painting stages. The company currently has a stock of raw materials for three months of production, as well as a small stock of semi-finished products. This strategy, adopted by Metal Cruz, aims to fulfil orders in a short space of time.

The layout of the company (figure 5) has not been planned in advance, so the machines and equipment are not arranged in such a way as to enable a flow of materials and products in a line. As for the activities within the production process, they are carried out by the only employee and the owner, who takes over the operation of all the equipment if necessary.

Due to the current high demand for production, the company plans to hire new staff at the beginning of 2016, so that it can serve its customers more quickly, as well as allowing for better planning of stocks of products and raw materials.

The company works with suppliers in the metropolitan region as well as locally, which is a strategy to get better prices and guarantee the availability of its inputs and raw materials, avoiding interruptions in the production process.

Figure 5 - Metal Cruz company *layout*

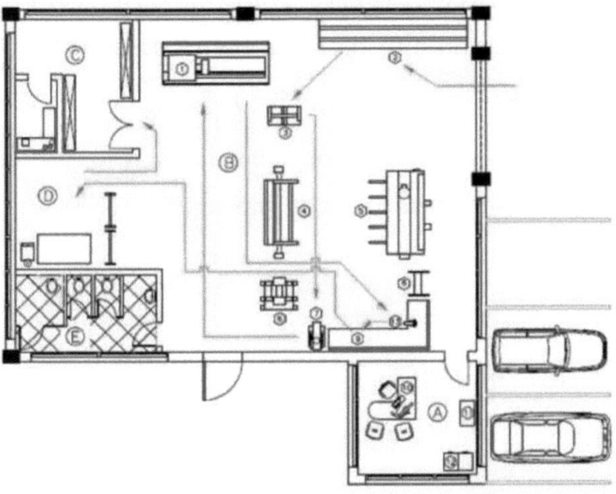

ARE AS	EQUIPMENT
A - Office	1 -Tome
B - Production	2 -Shelf for storing raw materials
C - Painting	3 - Sena polycutting
D - Welding	4 - Folding machine
E - Bathroom	5 - Guillotine
	6 - Calender
	7 -Bench drill
	8 - Hydraulic press
	9 - Bench
	10 -Office table
	11 -Table
	12 - Archive
	13 - Walrus

4.2 P+L implementation process.

The implementation of P+L in this study follows the CNTL guidelines, which divide this process into four distinct stages to be followed. Figure 6 shows the implementation stages suggested for the Metal Cruz company.

Stage 1: This stage sought the commitment of the company owner, making clear the objective of the methodology and the possible benefits, identifying barriers to implementation, defining the members of the ecotime, i.e. the group of people responsible for coordinating the implementation and maintenance of the methodology and the initial scope of P+L.

Stage 2: In the second stage, the flow of raw materials and the inputs and outputs of each activity in the production process were defined. Waste was then classified and a survey was carried out to visualise and quantify the raw materials used and the waste generated at each stage of the process.

Stage 3: Stage 3 defines the causes of waste generation according to the ecotime assessment. In this stage, a proposal for indicators for the production process was also drawn up.

Stage 4: In stage 4, the P+L opportunities were developed by ecotime.

Figura 6 - Stages of the P+L implementation process

ETAPA 1
Comprometimento da empresa;
Identificação de barreiras;
Formação do ecotime;
Estudo de abrangência;

ETAPA 2
Desenvolvimento do fluxograma do processo produtivo;
Classificação dos resíduos;
Levantamento de dados quantitativos;

ETAPA 3
Identificação das causas de geração de resíduos;
Proposta de indicadores;

ETAPA 4
Identificação das oportunidades de P+L;

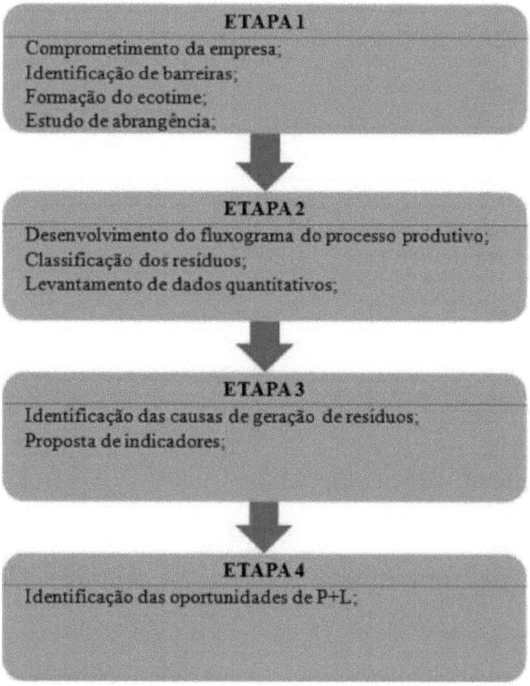

4.2.1 Step 1

The concepts relating to the P+L methodology were presented to the owner of the company under study, making it clear what its objectives are, the possible benefits that the methodology can bring to the production process and the barriers that can hinder the implementation process, obtaining his approval and support in the development of a P+L implementation proposal for Metal Cruz. Next, the barriers were defined, specific to the company's condition, to be overcome through the strategies established by ecotime. They are:

- Low availability of financial resources;
- Small number of employees, which makes it difficult for them to get involved in activities unrelated to the company's core process;
- High production demand;
- No waste management system;
- Technological limitations;
- Lack of resources needed for data collection.

Due to the small number of people involved in the company's activities, the ecotime is made up of the owner, the company's only employee and the author of this study, thus meeting the CNTL's recommendation of a minimum of three people and a maximum of ten to make up the ecotime. Taking into account the fact that the author is not an employee of the company and that he will leave the ecotime at the end of this study, it is recommended that the ecotime continue with just two members, until more employees are hired and at least one of them joins the group. As for the scope of P+L within the company, the manufacturing processes for the 12 products supplied to the client Xalingo S/A were chosen, as they are the company's main products and have well-defined processes, although they are not very standardised and always follow the same flow.

4.2.2 Stage 2

The production process flowchart was developed at this stage and can be found in figure 7, where you can also see all the inputs and outputs identified at each stage, allowing you to clearly visualise the waste generated.

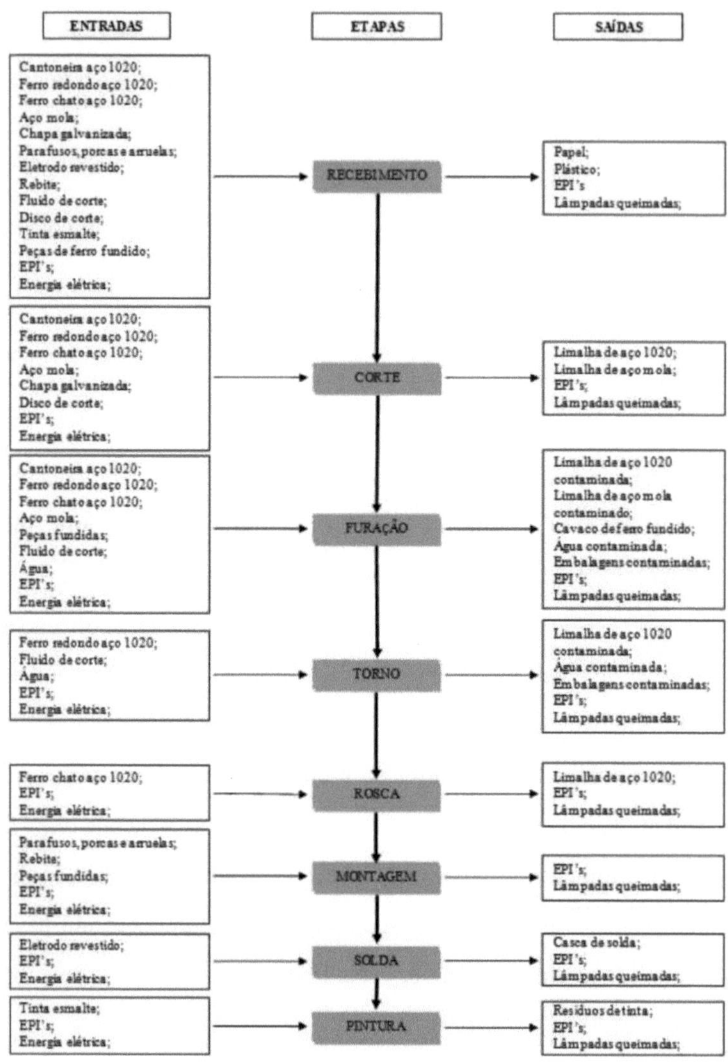

Figura 7 - Flowchart of process inputs and outputs

After establishing the input-output flowchart, table 2 was developed, which lists the equipment used in the production process and its energy consumption. Table 2 was used as the basis for surveying the energy inputs.

Table 2 - List of equipment and energy consumption

EQUIPMENT	POWER	CONSUMPTION (KWh/h)
Lathe	1/3 (hp)	0,245
Bench drill	^ (cv)	0,368
Saw (Policorte)	1 (cv)	0,736
Electrode welding	11500 (W)	11,5
Compressor	1 (cv)	0,736
Lamps	40 (W)	0,04

Table 3 shows the classification of each item of waste generated, according to ABNT standard NBR 10004, and Table 3 shows the data relating to the survey of process inputs, where the values correspond to the monthly average for the period from June 2014 to June 2015. For electrical energy inputs, the value indicated for each stage corresponds to the sum of the consumption of the equipment and the lamps used.

Chart 3 - Classification of waste generated by Metal Cruz

WASTE	CLASS
Paper	Class IIA
Plastic	Class IIA
1020 and spring steel filings	Class IIA
Contaminated 1020 steel filings	Class I
Contaminated spring steel filings	Class I
Cast iron chip	Class IIA
Abrasive powder	Class IIA
Welding shell	Class IIA
Paint residue	Class I
Contaminated packaging	Class I
PPE used	Class IIA
Burnt-out light bulbs	Class I
Contaminated water	Class I

Table 3 - Survey of process inputs

ENTRIES	QUANTITIES/MONTH							
	Receive.	Cutting	Drilling	Lathe	Thread	Assembly	Welding	Painting
Steel angle bracket 1020 (m)	6,75	6,75	6,75	-	-	-	-	-

1020 steel round iron (m)	29,22	22,3	12,8	17,7	-	-	-	-
1020 steel flat iron (m)	39,95	39,95	39,95	-	12,5	-	-	-
Spring steel (m)	12,5	12,5	12,5	-	-	-	-	-
Screws, nuts and washers (pcs)	4359	-	-	-	-	4359	-	-
Rivet (pcs)	248	-	-	-	-	248	-	-
Enamel paint (Kg)	3,84	-	-	-	-	-	-	3,84
Cutting fluid (l)	0,86	-	0,5	0,36	-	-	-	-
Cutting disc (pcs.)	23	23	-	-	-	-	-	-
Coated electrode (pcs)	46	-	-	-	-	-	46	-
Cast iron parts (pcs)	121	-	121	-	-	121	-	-
Galvanised sheet (unit)	0,002	0,002	-	-	-	-	-	-
PPE	4	2	2	2	1	1	2	1
Electricity (KWh/month)	7,04	341,75	236,91	113,51	7,04	7,04	409,66	49,97
Water (l)	-	-	10	7,33	-	-	-	-

Table 4 shows the outputs identified in Metal Cruz's production process, also taking into account the monthly average of waste generated in each of the production stages for the period from June 2014 to June 2015. For paint waste, an average loss of 40% was considered, bearing in mind that manufacturers establish losses of between 30 and 50% for the conventional spray painting process.

Table 4 - Quantification of process outputs

OUTPUTS	QUANTITIES PER PROCESS

	Receive.	Cutting	Drilling	Lathe	Thread	Assembly	Welding	Painting
Paper (Kg)	0,92	-	-	-	-	-	-	-
Plastic (Kg)	0,6	-	-	-	-	-	-	-
1020 steel filings (Kg)	-	1,72	-	-	0,003	-	-	-
Contaminated 1020 steel filings (Kg)	-	-	0,62	0,47	-	-	-	-
Contaminated spring steel filings (Kg)	-	0,2	0,04	-	-	-	-	-
Cast iron chips (Kg)	-	-	4,95	-	-	-	-	-
Welding shell (Kg)	-	-	-	-	-	-	0,033	-
Paint residue	-	-	-	-	-	-	-	1,54
Contaminated packaging (units)	-	-	1	1	-	-	-	-
PPE used (units)	4	2	2	2	1	1	2	1
Light bulbs burnt (units)	1	1	1	1	1	1	1	1
Water contaminated (l)	-	-	9,92	8,79	-	-	-	-

4.2.3 Stage 3

With the data collected, ecotime assessed the causes of waste generation in the company. Figure 8 shows these causes.

Figura 8 - Causes of waste generation in Metal Cruz's production process

OPERATIONAL	Old and poorly maintained lighting system: Machines with the main switch switched on unnecessarily: Steps carried out without a standard definition: Poorly dimensioned machines.

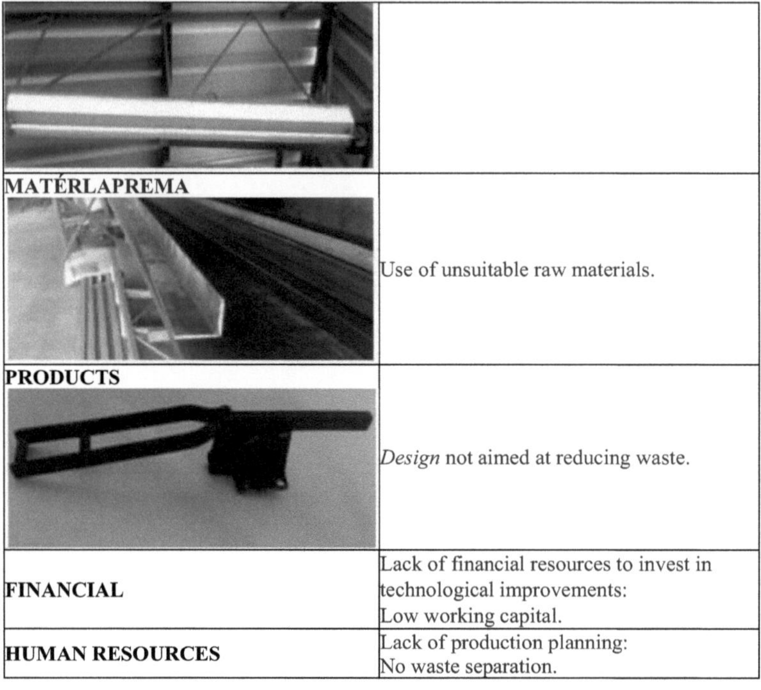

MATÉRLAPREMA	Use of unsuitable raw materials.
PRODUCTS	*Design* not aimed at reducing waste.
FINANCIAL	Lack of financial resources to invest in technological improvements: Low working capital.
HUMAN RESOURCES	Lack of production planning: No waste separation.

It can be seen that many of the causes of waste are linked to the use of electrical equipment, which makes energy losses an important point in the development of P+L opportunities. Even in cases where the cause is not directly related to the use of electrical equipment, it is possible, through improvements, to achieve a reduction in energy consumption by reducing the time taken to carry out the steps.

Once the causes had been established, a proposal for environmental performance indicators was developed, as shown in Table 4. The choice of indicators took into account the relevance of the information they can generate to help improve environmental issues and their ease of measurement, given the lack of human resources available to control the data and extract the information.

Table 4 - Proposed indicators

ENVIRONMENTAL ASPECT	INDICATOR
Energy consumption	KWh/Unit produced
Water consumption	m^3/Number of employees
Raw material consumption	m/Unit produced
Waste generation	Kg/Unit produced
Recycled waste	Kg of waste recycled/Kg of waste generated
Environmental objective	%improvements made/year

4.2.4 Stage 4

In the fourth stage, ecotime carried out a detailed analysis of the causes of waste generation identified in this study, together with direct monitoring of the production process, in order to identify P+L opportunities. This subchapter presents all the opportunities identified, as shown in Table 5.

Table 5 - P+L opportunities

LEVEL	OPPORTUNITY
Level 1	Replacement of the lighting system
	Adoption of measures to reduce water consumption Substitution of raw materials
	Technology replacement
	Change in the production process
	Creating operating procedures
Level 3	Implementation of a waste separation system

4.2.4.1 Replacing the lighting system

The current lighting system is made up of 40W fluorescent lamps. These bulbs contain components that are harmful to the environment and are classified after use as class I solid waste. In addition, they have a much shorter lifespan than more modern lighting systems and higher consumption.

The ecotime proposal consists of replacing fluorescent lamps with LED (Light Emitter Diode) lamps, figure 9. This change is expected to reduce lighting energy consumption by 85 per cent, as well as reducing the generation of this waste.

Figura 9 - LED lamp

Source: www.flc.com.br.

Information posters about the need and importance of reducing electricity consumption will also be put up near the switches to raise awareness (figure 10).

Figura 10 - Information poster on reducing energy consumption

4.2.4.2 Adopting measures to reduce water consumption

Metal Cruz uses a very small amount of water in its production process . This resource is only used to dilute cutting fluid, but there is consumption in the company's toilets, so ecotime identified the possibility of implementing taps with automatic shut-off and urinals. These products help to reduce water waste, generating savings of up to 70 per cent compared to traditional systems, according to data from the main manufacturers. Although this measure did not initially lead to a significant reduction due to the company's low consumption, it was considered feasible by ecotime due to the importance of this natural resource and the company's growth forecast, which is expected to lead to new hires, thus increasing consumption. An information poster emphasising the importance of saving water was also developed and can be seen in figure 11.

Figure 11 - Information poster on reducing water consumption

4.2.4.3 Substitution of raw materials

Three products manufactured by Metal Cruz use 2"x2" 1020 steel angles as one of their main raw materials. In the current manufacturing process, a longitudinal cut is made on each piece of angle to make it measure 2"xl" according to the product design. This cutting process is one of the main generators of waste in the production process, with a quarter of the raw material becoming waste. The P+L opportunity found for this process is to replace 2"x2" angles with a 4"xl" "U" profile. The aim is to cut the profile in the centre of its A measurement, as shown in figure 12, resulting in two angles with one side measuring 1" and the other measuring 2" minus 1.5mm or half the thickness of the cutting disc. As can be seen, there is a need for a small adjustment to the design, reducing one side of the angle bracket by 1.5mm, but this change does not affect the functionality of the products.

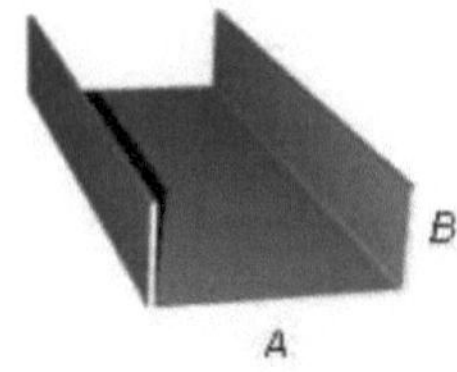

Figure 12 - U profile 4"xl"

With this substitution of raw materials, the company reduces the generation of waste in this case to close to zero and also reduces the time taken to use the saw by 50 per cent, as each cut generates two pieces, unlike the current process in which each piece needs to be cut individually. Energy consumption at this stage will also be reduced by 50 per cent.

Another important point to make the P+L opportunity viable is the economic factor, which in this case increases the benefit to the company. An angle bar six metres long costs around R$100.00 and generates around 75 pieces, depending on the product to be manufactured. A "U" profile bar, also six metres long, costs R$160.00 and generates twice as many parts.

4.2.4.4 Technology replacement

The cutting activity in Metal Cruz's production process is carried out using a polycutting saw. This equipment has a 1hp motor, i.e. a motor with a power of 736W or 0.736KW. This means that, when running at full load, it consumes 0.736 KWh/h (figure 13). ecotime identified the opportunity to replace this equipment with a band saw with a ^ hp motor, thus reducing consumption to 0.37 KWh/h (figure 14). This technological change also allows greater flexibility for the operator, who can put the saw into operation and dedicate himself to another activity in the process, returning only to remove the pieces already cut and position new ones. In this way, the company gains time when carrying out this activity, reducing production costs.

Fonte: Arquivo da empresa, 2015.

Figure 13 - Polycutting saw

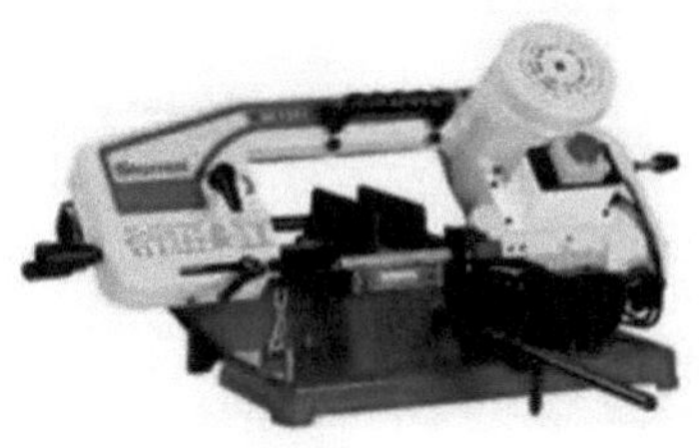

Fonte: www.leroymerlin.com.br.

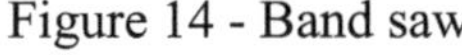

Figure 14 - Band saw

4.2.4.5 Change in the production process

In the current process, all the clamps manufactured use angle steel as one of the main raw materials, the size of which depends on the product manufactured. A polycut saw is used to cut this raw material (figure 13).

According to ecotime's assessment, the possibility of using flat iron bars to replace some of the sizes of angle brackets used was identified. In this way, the pieces would be cut on a guillotine and then bent on a bending machine. With this measure, the company avoids the generation of steel filings from disc cutting on the polycut saw. The opportunity found is easy to implement, as Metal Cruz already has this equipment, and all that is needed is the purchase of raw materials in the correct dimensions. This opportunity could be implemented to replace all the angle brackets for which there are commercially available flat iron measures that, when bent, produce angle brackets with similar-sized tabs.

4.2.4.6 Creating operating procedures

The lack of operational procedures within Metal Cruz's production process leads to waste, especially in terms of electricity, as an activity is often started without all the necessary material being available, causing unnecessary time spent with equipment switched on. The aim is to standardise tasks through operating procedures, optimising the use of equipment and reducing manufacturing time.

4.2.4.7 Implementation of a waste separation system

Metal Cruz does not currently separate its waste and has no control over its final disposal. The P+L opportunity found by ecotime provides for the correct separation of the waste generated, using containers (figure 15), adding value to it and making it possible to sell some of it to be used as raw material in other companies. For the other waste, the plan is to hire a company specialised in collection and final disposal. In this way, Metal Cruz will be able to monitor how its waste is disposed of by the contractor, reducing the environmental impact of its activities and complying with legislation, avoiding fines and other sanctions provided for by law.

Figure 15 - Rubbish bins

Source: www.maisplastico.com.br.

CHAPTER 5

CONCLUSION

This study aimed to identify P+L opportunities within the production process of a company in the metal-mechanics sector. It was motivated by the satisfactory results that the methodology has been achieving, by the activity carried out in the company that generates class I waste and by the lack of studies on this methodology.

The research presented a description of the main stages of P+L, structured in such a way as to fit in with Metal Cruz's production process. An investigation into the main causes of the company's waste generation, in which it was possible to realise that energy consumption is one of the biggest wastes, associated with a lack of planning in both the production process and the administrative structure. It also developed a proposal for environmental performance indicators to measure the company's progress with the implementation of P+L without presenting a degree of complexity in generating and analysing data that would make it unfeasible to use. Finally, it highlighted opportunities for improvements to be implemented to make Metal Cruz's activities less harmful to the environment, following the P+L priority levels.

It was clear that the application of P+L practices within the metal-mechanics sector generates advantages not only of an environmental nature, such as reducing damage to the environment and bringing the company into line with legislation, but also of a financial nature, reducing production costs, increasing efficiency and competitiveness, and even helping to improve the company's image in the eyes of its customers. In this way, this study proves that despite having simple solutions, P+L is an efficient and effective methodology that can bring numerous advantages to companies.

5.1 Suggestions for future work

For future studies, it is proposed to analyse and quantify the financial gains obtained by implementing P+L in the company, making it possible to extract data that will serve to motivate other companies to use this methodology. Studies that show the financial advantages can help overcome some of the main barriers to adopting P+L, especially those of an economic, organisational and technical nature.

REFERENCES

ALI, Y.; FRESNER, J. Half Is Enough - an Introduction to Cleaner Production. LCPC Press, Beirut, Lebanon, 2006.

AMATO NETO, João. Sustainability & Production: theory and practice for sustainable management. São Paulo: Atlas, 2011.

BRAZILIAN ASSOCIATION OF TECHNICAL STANDARDS. NBR 10004: Solid waste - classification. Rio de Janeiro, 2004.

BRAZILIAN ASSOCIATION OF TECHNICAL STANDARDS. NBR ISO 14001: Environmental Management System - Requirements with guidelines for use. Rio de Janeiro, 2004.

BAAS, Leo. To make zero emissions technologies and strategies become a reality, the lessons learned of cleaner production dissemination have to be known. Journal of Cleaner Production. v. 13, p. 1205-1216, 2007.

BRAGA, B. et al. Introduction to environmental engineering. São Paulo: Pearson Prentice Hall, 2002.

BROWN, L. R. et al. State of the World, 2001: Worldwatch Institute report on progress towards a sustainable society. Salvador: Ed., 2000.

CAMPOS, L. M. S.; MELO, D. A. Performance indicators of environmental management systems (EMS): a theoretical research. Production Vol. 18, n. 3, p 540-555, 2008.

CARDOSO, Lígia Maria França. Clean production indicators: A proposal for analysing companies' environmental reports. Dissertation (Professional Master's Degree in Environmental Management and Technology in the Production Process - Polytechnic School, Federal University of Bahia, Salvador, 2004.

CERVO, A. L.; BERVIAN, P. A.; SILVA, R. Metodologia científica. 6 ed. São Paulo: Pearson Prentice Hall, 2007.

CETESB. Technical environmental guide paints and varnishes - P+L series, 2008. Available at: http://www.cetesb.sp.gov.br/tecnologia/producao_limpa/documentos/tintas.pdf. Accessed on: 03 May 2015.

CNTL, National Centre for Clean Technologies. Implementation of Cleaner Production Programmes. National Centre for Clean Technologies SENAI-RS/UNIDO/UNEP, 42p., Porto Alegre, 2003.

CNTL, National Centre for Clean Technologies. What is the advantage of adopting Cleaner Production? Porto Alegre: CNTL, 2015a Available at: <http://www.senairs.com.br/cntl/>. Accessed on: 30 March 15.

COELHO, Arlinda Conceição Dias. Evaluation of the application of the UNIDO/UNEP Cleaner Production methodology in the sanitation sector - case study: EMBASA S. A. Dissertation (Professional Master's Degree in Environmental Management and Technologies in the Production Process) - Polytechnic School, Federal University of Bahia, Salvador, 2004.

DIAS, Patrícia. Co-operative actions between client and supplier companies to obtain socio-environmental benefits: A multiple case study in the metal-mechanic sector. Dissertation (Master's in Business Administration) - Federal University of Rio Grande do Sul, Porto Alegre, 2008.

DIAS, Reinaldo. Environmental management: social responsibility and sustainability. 2 ed.

DONG, X. et al. Application of a system dynamics approach for assessment of the impact of regulations on cleaner production in the electroplating industry in China. Journal of Cleaner Production, 2012.

FLC. Light bulbs. Available at: <http://www.flc.com.br/br/produtos/aplicacao/industrial>. Accessed on: 04 November. 2015.

GIL, Antonio Carlos. How to prepare research projects. 5 ed. São Paulo: Atlas, 2010.

HAWKEN, P. et al. Natural capitalism. Original title: Natural capitalism. [S.l.:s.n], 1999.

HIRSCHHORN, J. S. Why the pollution prevention revolution failed and why it ultimately will succeed. Pollut. 1997. Available at: < http://dx.doi.org/10.1002/(SICI)1520-6815(199724)7:1<11::AID-PPR2>3.0.CO;2-B.>. Accessed on: 27 March 2015.

KITZBERGER, J. et al. Adoption of sustainable production processes in a galvanising company In: XIX Simpósio de Engenharia de Produção, Anais... Bauru, 2012.

KLEMES, J. J.; VARBANOV, P. S. HUISINGH, D. Recent cleaner production advances in process monitoring and optimisation. Journal of Cleaner Production, 2012.

LEROY MERLIN. Building materials network. Available at: <http://www.leroymerlin.com.br/serras-de-corte-metal>. Accessed on: 03 November. 2015.

LIMA, L. B. et al. Survey and evaluation of environmental aspects and impacts in a metal-mechanic industry based on ISO 14000. In: XXXIV Encontro Nacional de Engenharia de Produção, Anais... Curitiba, 2014a.

LIMA, Y. C. C. et al. Lean construction and P+L as a quality management tool in the construction industry: A competitive strategy. In: XXXIV Encontro Nacional de Engenharia de Produção, Anais... Curitiba, 2014b.

MORE PLASTICS. Electronic market. Available at: <http://www.maisplastico.com.br/detalhes-produto.php?codigo=124081>. Accessed on: 03 November. 2015.

MALHOTRA, N. K. et al. Introduction to marketing research. São Paulo: Pearson Prentice Hall, 2005.

MARCONI, M. A.; LAKATOS, E. M. Técnicas de Pesquisa. 7 ed. São Paulo: Atlas, 2008.

MITCHELL, Gordon. Problems and Fundamentals of sustainable development indicators. 2004. Available at: http://www.lec.leeds.ac.uk/people/gordon. html. Accessed on 25 May. 2015.

OLIVEIRA, J. A. et al. Identification of the benefits and difficulties of cleaner production in industrial companies in the state of São Paulo. In: XXXIII Encontro Nacional de Engenharia de Produção, Anais... Curitiba, 2013.

PEREIRA, A. C.; SILVA, G. Z.; CARBONARI, M. E. E. Sustainability, social responsibility and the environment. São Paulo: Saraiva, 2012.

ROMM, J. J. Eco-efficient companies: How companies increase productivity and profits by reducing pollutant emissions. São Paulo: Signus, 2004.

SANTOS, Antônio Raimundo dos. Metodologia científica: a construção do conhecimento - 3 ed. - Rio de Janeiro: DP&A editora, 2000.

SELENE, R. et al. Creativity and innovation in negotiations with suppliers: Metal-mechanic industry in Paraná and São Paulo (Case study) In: XXXIV Encontro Nacional de Engenharia de Produção, Anais... Curitiba, 2014.

SEVERO, E. A. et al. Cleaner production, environmental sustainability and organisational performance: an empirical study in the Brazilian Metal-Mechanic industry. Journal of Cleaner Production, 2014.

SHI, H. et al. Barriers to the implementation of cleaner production in Chinese SMEs: govemment, industry and expert stakeholders' perspectives. Journal of Cleaner Production, 2007.

SILVA, A. E. et al. Application of the P+L methodology to reduce waste in tobacco processing companies. Tecno-Lógica Vol. 18, n. 2, p 97102, 2014a.

SILVA, A. E. et al. Cleaner production as a tool for competitive advantage in small companies: An overview and analysis of the main points of the tool. In: XXI Production Engineering Symposium, Proceedings...Bauru, 2014b.

TAN, X. C. et al. A decision-making framework model od cutting fluid selection for green manufacturing and a case study. Journal of Materials Processing Technology 129, 467-470. China. 2002.

UNEP and WBCSD; Cleaner Production and Eco-efficiency, Complementary Approaches to Sustainable Development, 1998.

UNEP (United Nations Environmental Programme), 2002. Available: <http://www.uneptie.org/pc/cp>. Accessed on: 01 June 2015.

WU, D. D.; OLSON, D. L.; BIRGE, J. R. Risk management in cleaner production. Journal of Cleaner Production, 2013.

YIN, R. K. Case study: Planning and methods. 5 ed. Porto Alegre: Bookman, 2015.

ZHANG, X. H. et al. The comparison of performances of a sewage treatment system before and after implementing the cleaner production measure. Journal of Cleaner Production, 2014.

Printed by Books on Demand GmbH, Norderstedt / Germany